HVAC for Beginners

The Ultimate Guide to Mastering Heating,
Ventilation, and Air Conditioning Systems

copyrighted@2024

Mark Harry

Table of Contents

Chapter One

HVAC Systems
Introduction to HVAC Systems

HVAC stands for Heating, Ventilation, and Air Conditioning, and it refers to technologies that control temperature, humidity, and air quality in indoor environments. HVAC systems are essential for maintaining comfort, safety and health in residential, commercial and industrial buildings. They ensure that the indoor environment remains within the desired range regardless of the external weather conditions. An efficient HVAC system is essential for providing a favorable environment for living, working and various industrial processes.

Significance and Application

HVAC systems are an integral part of various settings due to their role in maintaining a comfortable and healthy indoor environment. Some key applications include:

1. Residential:

Ensure comfortable living conditions by regulating indoor temperature and humidity.

Improve indoor air quality by filtering and purifying the air, reducing allergens and mitigating the effects of pollution.

2. Commercial:

Provide a pleasant environment for employees and customers, increase productivity and satisfaction.

Protect sensitive equipment and data centers from overheating and moisture damage.

3. Industrial:

Maintain optimal conditions for production processes, storage and equipment operation.

Ensure safety and compliance with health and safety standards at work.

4. Health care:

Create sterile and controlled environments in hospitals, clinics and laboratories to prevent the spread of infections and ensure patient comfort.

5. Education:

Provide a conducive environment for learning and teaching by maintaining comfortable temperatures and good air quality in schools and universities.

Basic Principles of HVAC

Thermodynamics and Heat Transfer

An understanding of thermodynamics and heat transfer is essential to the design and operation of HVAC systems. Thermodynamics is the study of energy, heat and their transformations, while heat transfer refers to the movement of heat from one place to another. There are three primary methods of heat transfer:

1. Conduction:

Heat transfer through a solid material.

It occurs when heat moves through a material from an area of high temperature to an area of low temperature.

Example: Heat passing through a metal rod.

2. Convection:

Heat transfer via a fluid (liquid or gas).

It occurs when a fluid flows over a surface or when there is a temperature difference in the fluid.

Example: Warm air rises and circulates in the room.

3. Radiation:

Heat transfer by electromagnetic waves.

It occurs without the need for a medium (can occur in a vacuum).

Example: Heat from the sun warming the Earth.

Fluid Mechanics

Fluid mechanics is the study of fluids (liquids and gases) and the forces that act on them. In HVAC systems, fluid mechanics is essential to understanding how air and refrigerants move through ducts, ducts, and equipment. Key concepts include:

1. Pressure:

The force generated by a fluid per unit area.

In HVAC systems, pressure differences drive airflow and refrigerant.

2. Flow rate:

The amount of fluid moving through a spot per unit time.

Measured in cubic feet per minute (CFM) for air and gallons per minute (GPM) for liquids.

3. Resistance:

Opposition to fluid flow in a system.

Caused by factors such as friction in ducts and ducts and

obstructions such as bends and fittings.

Refrigeration Cycle

The refrigeration cycle is a fundamental process in HVAC systems that is used to cool spaces. It involves the transfer of heat from one area to another using a refrigerant. The cycle consists of four major parts:

1. Compressor:

It raises the pressure and temperature of the coolant. The refrigerant enters the compressor under low pressure and exits under high pressure.

2. Condenser:

It releases heat from the refrigerant to the surroundings.

The refrigerant enters the condenser as a high-pressure gas and exits as a high-pressure liquid.

3. Expansion valve:

It lowers the pressure and temperature of the coolant.

Refrigerant enters the expansion valve as a high-pressure liquid and exits as a low-pressure liquid.

4. Evaporator:

It absorbs heat from the space being cooled.

The refrigerant enters the evaporator as a low-pressure liquid and exits as a low-pressure gas.

Psychrometrics

Psychrometrics is the study of the properties of air-water vapor mixtures. It is critical to understanding and managing humidity levels in HVAC systems. Key concepts include:

1. Dry bulb temperature:

Air temperature measured with a regular thermometer.

Does not take air humidity into account.

2. Wet-bulb temperature:

Air temperature measured with a wet-bulb thermometer.

Indicates how much moisture is in the air.

3. Relative humidity:

The ratio of the current amount of moisture in the air to the maximum amount it can hold at a given temperature.

Expressed as a percentage.

4. Dew point:

The temperature at which air gets saturated with moisture and water vapor starts to condense.

Air Distribution

Proper air distribution is essential to maintain comfort and indoor air quality. This includes the design and arrangement of ducts, vents and diffusers to ensure even distribution of conditioned air. Key concepts include:

1. Duct design:

Duct design and sizing to minimize resistance and ensure efficient airflow.

Factors include duct material, shape and insulation.

2. Ventilation:

The process of supplying and removing air from a space to control air quality and comfort.

It can be achieved by natural or mechanical means.

3. Diffusers and grilles:

Devices used to distribute and direct air flow in a space.

Diffusers spread the air in different directions, while grilles control the flow direction.

Components of HVAC Systems

HVAC systems contain several key components that work together to regulate indoor temperature, humidity and air quality. This will explore the main components of heating, ventilation and air conditioning systems, detailing their functions and how they interact within an HVAC system.

Heating Systems

1. Furnaces:

Furnaces are heaters that produce heat by burning fuel (natural gas, oil, propane) or electrical resistance.

Components include burners, heat exchangers, blowers and controls.

Furnaces distribute heated air through ducts in a forced air system.

2. Boilers:

Boilers produce hot water or steam for heating by burning fuel or electricity.

Heated water or steam circulates through radiators or under-floor heating systems.

Components include burners, heat exchangers, circulation pumps and controls.

3. Heat pumps:

Heat pumps transfer heat from one place to another and provide both heating and cooling.

They work in a refrigeration cycle and use electricity to transfer heat.

Types include air source, ground source (geothermal) and water source heat pumps.

Ventilation Systems

1. Natural ventilation:

It uses natural forces such as wind and buoyancy to circulate the air.

Achieved through operable windows, vents and architectural design.

Benefits include energy savings and better indoor air quality.

2. Mechanical ventilation:

It uses fans and ducts to control air movement.

Types include exhaust, supply, balanced ventilation and energy recovery ventilation (ERV/HRV) systems.

Ensures consistent air exchange and filtration.

Air Conditioning Systems

1. Split systems:

It consists of an outdoor condensing unit and an indoor evaporator.

The refrigerant absorbs heat inside and releases it outside.
Common in both residential and modest business structures.

2. Central AC systems:

Use a network of ducts to distribute cooled air from the central unit.

A central unit usually contains a compressor, a condenser and an evaporator.

Suitable for large residential and commercial buildings.

3. Ductless Mini-Splits:

They comprise of an outside unit and one or more inside units linked by a refrigerant line.

Each indoor unit serves a certain zone or room.

It offers flexibility, energy efficiency and easy installation.

Air Distribution Components

1. Duct-work:

Ducts transport conditioned air throughout the building.

Made from materials such as galvanized steel, aluminum or flexible plastic.

Designed to minimize resistance and leakage.

2. Vents and registers:

Vents allow air to enter and exit the duct.

Registers are adjustable grilles that control the direction and volume of air flow.

Correct placement and size ensure efficient air distribution.

3. Diffusers:

Distribute the air evenly over the area.

Types include ceiling, wall and floor diffusers.

Designed to minimize drafts and noise.

Refrigeration Components

1. Compressor:

It raises the pressure and temperature of the coolant.

Necessary for the refrigeration cycle.

It is located in the outdoor unit of the air conditioning and heat pump systems.

2. Condenser coil:

It releases heat from the refrigerant to the surroundings.

Located in the outdoor unit.

Heat is dissipated through fins and fans.

3. Expansion valve:

It reduces the pressure and temperature of the coolant.

It regulates the flow of refrigerant into the evaporator.

It ensures efficient heat absorption.

4. Evaporator coil:

Absorbs heat from indoor air.

It is positioned within the indoor unit.

Cooled air is circulated by a blower or fan.

Control Systems

1. Thermostats:

Devices that regulate temperature by controlling HVAC components.

Types include manual, programmable and smart thermostats.

Smart thermostats provide remote control and learning capabilities.

2. Building Management Systems (BMS):

Centralized systems that monitor and control HVAC, lighting, security and other building functions.

Improve efficiency and comfort through automated management and data analysis.

3. Smart HVAC systems:

Use IoT (Internet of Things) technology for advanced control and monitoring.

It includes features such as remote access, energy consumption monitoring and predictive maintenance.

Increase convenience,
efficiency and comfort.

Chapter Two

Design and Installation of HVAC Systems

Designing and installing HVAC systems requires careful planning, accurate calculations and an understanding of building requirements. This will cover the fundamental aspects of HVAC system design and installation, including load calculation, system design and layout, ductwork design, and equipment selection and sizing.

Load Calculation

1. Calculation of heat load:

It determines the amount of heat needed to maintain internal comfort.

Factors include building size, insulation, window orientation and occupancy.

Calculated using Manual J (Residential) or Manual N (Commercial) standards.

2. Calculation of the cooling load:

It determines the amount of cooling required to maintain indoor comfort.

Factors include external heat gains (solar radiation, outside air) and internal heat gains (people, equipment).

Calculated using Manual J (Residential) or Manual N (Commercial) standards.

3. Software Tools:

Various software tools such as Carrier HAP, Trane TRACE and Wright-soft assist in accurate load calculations.

These tools take into account detailed building parameters and weather data for accurate results.

System Design and Layout

1. Zoning:

It divides the building into different areas, each with its own temperature control.

Improves comfort and energy efficiency by targeting specific areas.

Zoning systems can be achieved using multiple thermostats and dampers.

2. Location of equipment:

Optimal placement of HVAC equipment ensures efficiency and availability for maintenance.

Outdoor units should be located in well-ventilated areas, away from obstacles.

Indoor units should be centrally located for even air distribution.

3. Piping and ductwork routing:

Correct routing of ducts and ducts minimizes pressure losses and energy consumption.

Avoid sharp turns and long drives to reduce drag.

Insulate ducts and ducts to prevent heat loss or gain.

Ductwork Design

1. Duct sizing:

Correct sizing ensures efficient air distribution and minimizes noise.

Undersized ductwork can lead to high velocity and noise, while oversized ductwork is inefficient.

Use tools like Ductulator for precise sizing based on airflow requirements.

2. Duct layout:

The arrangement should minimize drag and ensure balanced air flow.

Main channels should have gradual transitions and minimal sharp turns.

Branch ducts should be appropriately sized and located for even distribution.

3. Material selection:

The duct can be made of galvanized steel, aluminum or flexible plastic.

Metal ducts are durable and less likely to leak, but require insulation.

Flexible ducting is easier to install, but should be used sparingly to avoid airflow problems.

Equipment Selection and Sizing

1. Heating and cooling equipment:

Select equipment based on load calculations and building requirements.

Consider efficiency ratings (SEER for cooling, AFUE for heating) to optimize energy use.

Ensure compatibility with existing systems or planned installations.

2. Air handlers and fans:

Choose air handlers and fans that match your system's capacity and airflow requirements.

Consider variable speed options for better efficiency and convenience.

Ensure proper sizing to avoid airflow and noise issues.

3. Ventilation equipment:

Select ventilation equipment based on the building's air exchange requirements.

For energy efficiency, consider energy recovery ventilators (ERVs) or heat recovery ventilators (HRVs).

Ensure proper filtration to maintain indoor air quality.

Installation Best Practices

1. Quality workmanship:

Ensure proper sealing of ducts and joints to prevent air leakage.

Use appropriate fasteners and supports for secure installation.

Installation should be performed in accordance with the manufacturer's instructions and industry standards.

2. Testing and Balancing:

Perform system testing to ensure proper function and performance.

Equalize airflow to achieve even distribution and prevent hot or cold spots.

For accurate measurements, use instruments such as manometers and anemometers.

3. Commissioning:

Commission the system and verify that it meets design specifications and operational requirements.

Document system performance, including airflow, temperature, and pressure data.

Provide training and documentation for building users or maintenance personnel.

Compliance and Standards

1. Building codes:

Ensure compliance with local, state and national building codes and regulations.

Obtain necessary permits and inspections during installation.

Follow safety standards for electrical, plumbing and design work.

2. Industry Standards:

Adhere to industry standards set by organizations such as ASHRAE (American Society for Heating, Refrigeration and Air Conditioning) and ACCA (Air Conditioning Contractors of America).

Implement best practices for energy efficiency, indoor air quality and system performance.

3. Environmental regulations:

Consider environmental regulations regarding refrigerants, emissions and energy consumption.

Use environmentally friendly refrigerants and high efficiency equipment to minimize impact.

Implement sustainable practices and technologies in system design and installation.

Control Systems and Automation

Control systems and automation are critical components of modern HVAC systems, enabling precise control of temperature, humidity and air quality. This examines the various control systems and automation technologies used in HVAC, including thermostats, building management systems, and intelligent HVAC systems.

Thermostats

1. Manual thermostats:

Basic thermostats that require manual adjustment to set the desired temperature.

Simple to use, but lacks advanced features and power saving options.

2. Programmable thermostats:

They allow users to set temperature change schedules

based on time of day and day of the week.

Help reduce energy use by adjusting temperatures when buildings are unoccupied or off-peak.

Features may include multiple programming periods, holiday modes and overrides.

3. Smart thermostats:

Connected thermostats that can be controlled remotely via smartphone apps or web interfaces.

Offer advanced features such as learning algorithms, geo-fencing, energy usage reports, and integration with other smart home devices.

Examples include Nest, Eco-bee, and Honeywell Lyric.

Building Management Systems (BMS)

1. Overview:

Centralized systems that monitor and control various building functions, including HVAC, lighting, security and fire safety.

Provide a unified interface to manage building operations, increase efficiency and convenience.

2. Components:

Sensors: Measures parameters such as temperature, humidity, occupancy and air quality.

Controllers: Process data from sensors and perform control actions to maintain desired conditions.

Actuators: Devices that adjust HVAC equipment based on control signals (e.g., valves, dampers, fans).

User interface: Software platforms that allow building operators to monitor and control the system often include graphical displays and reporting tools.

3. Advantages:

Energy efficiency: Optimize energy use by controlling HVAC and other systems based on real-time data and schedules.

Improved Comfort: Maintain consistent indoor conditions by responding to changes in occupancy and environmental factors.

Maintenance and diagnostics: Monitor system performance and identify problems early, reducing downtime and repair costs.

Smart HVAC Systems

1. Internet of Things (IoT):

IoT technology connects HVAC components to the Internet and enables advanced monitoring and control.

Sensors and devices transmit data to cloud platforms for analysis and optimization.

2. Remote monitoring and control:

Access and control HVAC systems from anywhere using mobile apps or web interfaces.

Receive alerts and notifications about system performance, maintenance needs, and faults.

3. Predictive maintenance:

Use data analytics and machine learning to predict equipment failures and maintenance needs.

Proactively schedule maintenance to avoid downtime and extend equipment life.

4. Energy management:

Monitor your energy consumption and look for areas where you might save money.

Implement demand response strategies to reduce energy costs during peak periods.

5. Integration with other systems:

Smart HVAC systems can be integrated with other building systems such as lighting, security and renewable energy sources.

It enables coordinated control and optimization for overall building performance.

Advanced Control Strategies

1. Variable Air Volume (VAV) systems:

Adjust the amount of air supplied to different zones as needed, increasing comfort and efficiency.

Use VAV boxes with flaps to control airflow to each zone.

2. Demand Controlled Ventilation (DCV):

Adjust ventilation rate based on occupancy levels measured by CO_2 sensors or occupancy sensors.

Reduce energy consumption by ventilating only when needed.

3. Zoning systems:

Divide buildings into zones with independent temperature control.

Use zone dampers and thermostats to optimize comfort and energy use in each area.

4. Advanced algorithms:

Implement algorithms for optimal control such as model predictive control (MPC) or fuzzy logic control.

Use real-time data and predictive models to make control decisions.

Human Machine Interface (HMI)

1. User Friendly Interfaces:

Design an intuitive interface for building operators and users to interact with HVAC systems.

Use graphic displays, touch screens and voice commands for ease of use.

2. Data visualization:

Present data in accessible formats such as dashboards, graphs, and reports.

Enable users to monitor system performance, power consumption and indoor conditions.

3. Customization and personalization:

Allow users to customize settings and preferences for convenience and efficiency.

Provide personalized recommendations for energy savings and system optimization.

Chapter Three

Maintenance and Troubleshooting

Proper maintenance and timely troubleshooting are key to optimal performance, energy efficiency and longevity of HVAC systems. This will cover the importance of regular maintenance, preventive maintenance practices, common problems and troubleshooting techniques for HVAC systems.

The Importance of Regular Maintenance

1. Efficiency:

Regular maintenance ensures that HVAC systems are operating at peak efficiency.

Clean and well-maintained components use less energy and lower operating costs.

2. Reliability:

Preventive maintenance helps identify and solve potential problems before they lead to system failure.

It increases the reliability of the system and reduces the probability of unexpected failures.

3. Lifespan:

Proper care and maintenance extends the life of your HVAC equipment.

Protects investment in HVAC systems by preventing premature wear.

4. Indoor air quality:

Regular cleaning and replacement of the filter maintains good indoor air quality.

Reduces the presence of allergens, dust and dirt.

5. Safety:

Ensures safe operation of HVAC systems, especially those involving combustion.

Prevents hazards such as carbon monoxide leaks and electrical faults.

Preventive Maintenance Procedures

1. Routine checks:

Perform periodic inspections of HVAC components for signs of wear, damage, or malfunction.

It includes visual inspections, functional tests and performance measurements.

2. Filter replacement:

Replace air filters regularly to ensure proper airflow and air quality.

The frequency depends on the type of filter and conditions of use, usually every 13 months.

3. Cleaning components:

Clean the fan coils, ducts and components to remove dust, dirt and debris.

Prevents airflow restriction and improves heat exchange efficiency.

4. Lubrication:

Lubricate moving parts such as bearings and motors to reduce friction and wear.

Use the manufacturer's recommended lubricants and follow the application instructions.

5. Checking the coolant level:

Make sure the coolant levels are within the range specified by the manufacturer.

A low refrigerant level can reduce the cooling efficiency and cause damage to the compressor.

6. Thermostat calibration:

Check and adjust thermostats to maintain proper temperature regulation.

Replace batteries in wireless thermostats as needed.

7. Check the electrical connections:

Check and tighten electrical connections to prevent loose wires and potential electrical malfunctions.

Check for signs of wear, corrosion or overheating.

Common HVAC Problems and Troubleshooting Techniques

1. Inadequate heating or cooling:

Check the thermostat settings: Ensure that the thermostat is set to the desired temperature and mode.

Check Air Filters: Dirty filters can restrict air flow and reduce system efficiency.

Examine the ducts: Look for leaks, blockages or disconnected ducts.

Check coolant levels: Low coolant levels can affect cooling performance.

2. The system does not turn on:

Power: Verify that the system has power and check circuit breakers or fuses.

Thermostat: Make sure the thermostat is working properly and has fresh batteries.

Safety switches: Check safety switches and sensors that may have tripped or failed.

3. Unusual sounds:

Loose parts: Check for loose parts such as screws, panels or ducts.

Motor and Bearings: Check motor bearings or fan components for wear or damage.

Debris: Look for debris caught in the fan or other moving parts.

4. Frequent cycling:

Thermostat location: Make sure the thermostat is not in a

location exposed to drafts or direct sunlight.

Oversized System: An oversized HVAC system can shorten the cycle; consult an expert for assessment.

Airflow problems: Check for blockages or restrictions in ducts and air filters.

5. Poor airflow:

Dirty filters: Replace or clean the air filters regularly.

Ductwork Leaks: Check and seal any ductwork leaks.

Fan problems: Check the fan motor and fan for proper operation.

6. Water leaks:

Condensate drain: Make sure the condensate drain is not blocked and drains properly.

Refrigerant Lines: Check refrigerant lines for insulation and potential leaks.

Evaporator Coil: Check the evaporator coil for ice build-up, which can indicate low refrigerant or airflow issues.

Advanced Troubleshooting and Diagnostic Tools

1. Multimeter:

   Use a multimeter to measure voltage, current, and resistance in electrical components.

   Essential for diagnosing electrical problems and ensuring safe operation.

2. Manometer:

   Measures air pressure in ducts and across filters to assess airflow and pressure drop.

   It helps identify blockages, leaks and system imbalances.

3. Refrigerant gauges:

Check system refrigerant pressure to ensure proper charge and operation.

Diagnose problems related to refrigerant levels and compressor performance.

4. Thermal imaging camera:

Detects temperature changes in HVAC components and ductwork.

Useful for identifying heat loss, insulation problems and component overheating.

5. Combustion analyzer:

It measures the efficiency and safety of combustion systems such as furnaces and boilers.

It ensures the correct fuel-air ratio and detects harmful emissions.

6. Data loggers:

Record temperature, humidity and other environmental conditions over time.

Provide an overview of system performance and help identify intermittent problems.

Chapter Four

Energy Efficiency and Sustainability in HVAC Systems

Energy efficiency and sustainability are increasingly important in the design, operation and maintenance of HVAC systems. It explores strategies, technologies and practices to increase the energy efficiency and sustainability of HVAC systems reduce environmental impact and operating costs.

The Importance of Energy Efficiency and Sustainability

1. Reduced operating costs:

Energy-efficient HVAC systems use less energy, leading to lower utility bills.

Investments in energy-efficient technologies can lead to significant long-term savings.

2. Impact on the environment:

Efficient systems reduce greenhouse gas emissions and rely on fossil fuels.

Sustainability initiatives contribute to mitigating climate change and conserving natural resources.

3. Regulatory compliance:

Compliance with energy efficiency standards and regulations helps avoid penalties and incentives.

Governments and organizations are increasingly mandating sustainability practices in building operations.

4. Higher comfort and performance:

Energy efficient systems often provide better temperature and humidity control.

Improved indoor air quality thanks to effective ventilation and filtration.

Strategy for Energy Efficiency

1. Highly efficient equipment:

Choose HVAC equipment with high seasonal energy efficiency (SEER) and annual fuel efficiency (AFUE) ratings.

Consider Energy Star certified products for guaranteed energy performance.

2. Variable speed technology:

Use variable speed drives (VSDs) for compressors, fans and pumps to match system capacity to requirements.

It reduces energy consumption at partial load and increases comfort.

3. Zoning systems:

Implement zoning to independently control heating and cooling in different areas.

It prevents the modification of unoccupied spaces and reduces energy waste.

4. Insulation and sealing:

Properly insulate ducts, ducts, and building envelopes to minimize heat loss or gain.

Seal leaks in ducts and around doors and windows to improve system efficiency.

5. Energy recovery ventilation (ERV):

Use ERV systems to recover energy from exhaust air and prepare fresh air.

Reduces the load on HVAC systems by recovering heat and moisture.

6. Demand Controlled Ventilation (DCV):

Adjust ventilation rate based on occupancy using CO_2 or occupancy sensors.

It ensures adequate ventilation while minimizing energy consumption when spaces are unoccupied.

7. Regular maintenance:

Perform routine maintenance to keep systems running at peak efficiency.

Clean and replace filters, check and clean exchangers, and check coolant levels.

Sustainable HVAC Practices

1. Integration of renewable energy:

Incorporate renewable energy sources such as solar, wind or

geothermal into your HVAC systems.

Solar panels can power HVAC components, while geothermal heat pumps use the Earth's stable temperature for heating and cooling.

2. Certification of eco-friendly buildings:

Focus on certifications such as LEED (Leadership in Energy and Environmental Design) or BREEAM (Building Research Establishment Environmental Assessment Method).

These certifications support sustainable practices and energy-efficient designs.

3. Sustainable materials:

Use environmentally friendly materials for insulation, ductwork and other HVAC components.

Recyclable and Non-Toxic Materials Contribute to Sustainability.

4. Life cycle assessment:

Consider the entire life cycle of HVAC systems, from production to disposal.

Choose products and systems with a lower environmental impact over their lifetime.

5. Smart control and automation:

Implement advanced control systems for optimized HVAC operation.

Use smart thermostats, building management systems (BMS) and IoT-enabled devices for real-time monitoring and control.

6. Water conservation:

Use water-saving technologies in HVAC systems, such as cooling towers with efficient water management practices.

Implement rainwater harvesting or gray-water reuse for cooling tower make-up water.

Emerging Technologies and Innovations

1. Advanced heat pumps:

Highly efficient heat pumps, including air, ground and water sources, offer sustainable heating and cooling options.

Innovations in heat pump technology continue to increase performance and efficiency.

2. Magnetic cooling:

An emerging technology that uses magnetic fields to achieve

cooling without traditional refrigerants.

It offers the potential for high efficiency and reduced environmental impact.

3. Absorption coolers:

Use heat sources such as natural gas, sunlight, or waste heat to drive the refrigeration cycle.

It can be an effective alternative to traditional electric chillers, especially in combined heat and power (CHP) systems.

4. Phase Change Materials (PCM):

PCMs store and release thermal energy during phase transitions (eg melting/freezing).

It is used in building materials and HVAC systems for thermal energy storage and load transfer.

5. Hybrid systems:

Combine multiple HVAC technologies to optimize performance and efficiency.

Examples include hybrid heat pumps or systems that integrate renewable energy sources with conventional HVAC equipment.

Chapter Five

Indoor Air Quality (IAQ)

Indoor air quality (IAQ) is a critical aspect of HVAC systems that directly affects the health, comfort and productivity of building occupants. It delves into factors affecting IAQ, common pollutants and strategies for improving and maintaining high indoor air quality.

The Importance of Indoor Air Quality

1. Health benefits:

   Poor IAQ can cause respiratory problems, allergies, and other health issues.

Ensuring good IAQ helps prevent disease and promotes overall well-being.

2. Convenience and Productivity:

Good IAQ increases passenger comfort, leading to higher productivity and satisfaction.

It reduces absenteeism at workplaces and improves the learning environment in schools.

3. Regulatory compliance:

Compliance with IAQ standards and guidelines ensures a safe and healthy indoor environment.

Organizations such as ASHRAE and the EPA provide standards for acceptable levels of IAQ.

Factors Affecting Indoor Air Quality

1. Ventilation:

Adequate ventilation is required to dilute and eliminate indoor contaminants.

Both natural and mechanical methods of ventilation can be used.

2. Filtration:

Air filters remove particles, allergens and other impurities from the air.

The efficiency of the filters is evaluated by the MERV value (Minimum Efficiency Reporting Value).

3. Humidity control:

Maintaining the correct humidity level (3060%) prevents mold growth and reduces dust mites.

Dehumidifiers and humidifiers can be used to regulate indoor humidity.

4. Source control:

Identifying and eliminating or reducing sources of indoor pollution.

Examples include the use of low-emission building materials, proper storage of chemicals, and management of combustion sources.

Common Indoor Air Pollutants

1. Particulate matter (PM):

Includes dust, pollen, smoke and other fine particles.

It can worsen breathing problems and cause other health problems.

2. Volatile Organic Compounds (VOCs):

Emitted products such as paints, cleaning products and building materials.

Long-term exposure to VOCs can lead to various health problems.

3. Biological contaminants:

Include mold, bacteria, viruses, and allergens from pets or pests.

It can cause infections, allergic reactions and other health problems.

4. Carbon dioxide (CO_2):

High CO_2 levels can indicate poor ventilation and lead to discomfort and reduced cognitive function.

Maintaining proper ventilation helps keep CO_2 levels under control.

5. Carbon Monoxide (CO):

Incomplete combustion produces a colorless and odorless gas.

High levels can be fatal; Proper ventilation and CO detectors are crucial.

6. Radon:

A naturally occurring radioactive gas that can leak into structures from the earth.

Long-term exposure increases the risk of lung cancer; testing and mitigation are essential.

Strategies for Improving Indoor Air Quality

1. Improved ventilation:

Increase the amount of outdoor air supplied to the building.

Use energy recovery ventilators (ERVs) or heat recovery ventilators (HRVs) to boost ventilation efficiency.

2. Air filtration:

Use high-efficiency particulate air (HEPA) filters to remove fine particles.

Replace or clean the filters regularly to maintain their effectiveness.

3. Humidity control:

Use dehumidifiers to reduce humidity levels in damp areas such as basements.

Use humidifiers in dry climates to maintain a comfortable humidity level.

4. Source control:

Choose low VOC products and materials.

Store chemicals and hazardous materials in well-ventilated areas or outside living quarters.

5. Air purification:

Use air purifiers with HEPA filters or activated carbon filters to remove pollutants.

Consider advanced air purification technologies such as UVC light and ionizers.

6. Regular maintenance:

Keep HVAC systems well maintained to ensure optimal performance.

Clean ducts, coils and other components regularly to prevent dirt build-up.

7. Monitoring and Testing:

With IAQ monitors, you can monitor pollutant levels, humidity and temperature.

Carry out regular tests for radon, CO and other harmful substances.

IAQ in Different Settings

1. Residential buildings:

Focus on ventilation, filtration and moisture control.

Focus on common sources of indoor pollution such as cooking, smoking and household chemicals.

2. Commercial buildings:

Implement advanced HVAC systems with high efficiency filters and ventilation controls.

Monitor IAQ regularly to ensure a healthy environment for passengers.

3. Schools and kindergartens:

Prioritize good IAQ to protect children's health and improve learning conditions.

Ensure proper ventilation and use materials and cleaning agents with low emissions.

4. Medical facilities:

Maintain strict IAQ standards to prevent the spread of infections.

Use advanced filtration and air cleaning systems along with regular IAQ monitoring.

Indoor Air Quality (IAQ)

Indoor air quality (IAQ) is a critical aspect of HVAC systems that directly affects the health, comfort and productivity of building occupants. It delves into factors affecting IAQ, common pollutants and strategies for improving and maintaining high indoor air quality.

The Importance of Indoor Air Quality

1. Health benefits:

Poor IAQ can cause respiratory problems, allergies, and other health issues.

Ensuring good IAQ helps prevent disease and promotes overall well-being.

2. Convenience and Productivity:

Good IAQ increases passenger comfort, leading to higher productivity and satisfaction.

It reduces absenteeism at workplaces and improves the learning environment in schools.

3. Regulatory compliance:

Compliance with IAQ standards and guidelines ensures a safe and healthy indoor environment.

Organizations such as ASHRAE and the EPA provide standards for acceptable levels of IAQ.

Factors Affecting Indoor Air Quality

1. Ventilation:

Adequate ventilation is required to dilute and eliminate indoor contaminants.

Both natural and mechanical methods of ventilation can be used.

2. Filtration:

Air filters remove particles, allergens and other impurities from the air.

The efficiency of the filters is evaluated by the MERV value (Minimum Efficiency Reporting Value).

3. Humidity control:

Maintaining the correct humidity level (3060%) prevents mold growth and reduces dust mites.

Dehumidifiers and humidifiers can be used to regulate indoor humidity.

4. Source control:

Identifying and eliminating or reducing sources of indoor pollution.

Examples include the use of low-emission building materials, proper storage of chemicals, and management of combustion sources.

Common Indoor Air Pollutants

1. Particulate matter (PM):

Includes dust, pollen, smoke and other fine particles.

It can worsen breathing problems and cause other health problems.

2. Volatile Organic Compounds (VOCs):

Emitted products such as paints, cleaning products and building materials.

Long-term exposure to VOCs can lead to various health problems.

3. Biological contaminants:

Include mold, bacteria, viruses, and allergens from pets or pests.

It can cause infections, allergic reactions and other health problems.

4. Carbon dioxide (CO_2):

High CO2 levels can indicate poor ventilation and lead to discomfort and reduced cognitive function.

Maintaining proper ventilation helps keep CO2 levels under control.

5. Carbon Monoxide (CO):

Incomplete combustion produces a colorless and odorless gas.

High levels can be fatal; Proper ventilation and CO detectors are crucial.

6. Radon:

A naturally occurring radioactive gas that can leak into structures from the earth.

Long-term exposure increases the risk of lung cancer; testing and mitigation are essential.

Strategies for Improving Indoor Air Quality

1. Improved ventilation:

Increase the amount of outdoor air supplied to the building.

Use energy recovery ventilators (ERVs) or heat recovery ventilators (HRVs) to boost ventilation efficiency.

2. Air filtration:

Use high-efficiency particulate air (HEPA) filters to remove fine particles.

Replace or clean the filters regularly to maintain their effectiveness.

3. Humidity control:

Use dehumidifiers to reduce humidity levels in damp areas such as basements.

Use humidifiers in dry climates to maintain a comfortable humidity level.

4. Source control:

Choose low VOC products and materials.

Store chemicals and hazardous materials in well-ventilated areas or outside living quarters.

5. Air purification:

Use air purifiers with HEPA filters or activated carbon filters to remove pollutants.

Consider advanced air purification technologies such as UVC light and ionizers.

6. Regular maintenance:

Keep HVAC systems well maintained to ensure optimal performance.

Clean ducts, coils and other components regularly to prevent dirt build-up.

7. Monitoring and Testing:

With IAQ monitors, you can monitor pollutant levels, humidity and temperature.

Carry out regular tests for radon, CO and other harmful substances.

IAQ in Different Settings

1. Residential buildings:

Focus on ventilation, filtration and moisture control.

Focus on common sources of indoor pollution such as cooking, smoking and household chemicals.

2. Commercial buildings:

Implement advanced HVAC systems with high efficiency filters and ventilation controls.

Monitor IAQ regularly to ensure a healthy environment for passengers.

3. Schools and kindergartens:

Prioritize good IAQ to protect children's health and improve learning conditions.

Ensure proper ventilation and use materials and cleaning agents with low emissions.

4. Medical facilities:

Maintain strict IAQ standards to prevent the spread of infections.

Use advanced filtration and air cleaning systems along with regular IAQ monitoring.

Chapter Six

Advances and Future Trends in HVAC

The HVAC industry is constantly evolving, driven by technological advances, changing regulatory requirements and a growing emphasis on sustainability. This explores the latest advancements and future trends in HVAC technology, highlighting the innovations that are shaping the future of HVAC.

Technological Advances in HVAC

1. Smart HVAC systems:

The integration of Internet of Things (IoT) technology enables HVAC systems to be connected, monitored and controlled remotely.

Features include predictive maintenance, real-time energy monitoring and adaptive control based on user behavior and environmental conditions.

2. Artificial Intelligence (AI) and Machine Learning:

Artificial intelligence algorithms analyze data from sensors and other sources to optimize HVAC performance.

Machine learning models predict system failures, optimize energy consumption and improve passenger comfort through intelligent control.

3. Variable Refrigerant Flow (VRF) systems:

VRF systems provide precise control of refrigerant flow to multiple indoor units, allowing individual temperature control in different zones.

Highly efficient and adaptable to different types and sizes of buildings.

4. Advanced heat pumps:

Development of highly efficient heat pumps that work effectively in colder climates.

Innovations include air, ground and water source heat pumps with improved performance and lower environmental impact.

5. Magnetic refrigeration:

Cutting-edge cooling technology that uses a magnetic field to achieve cooling without traditional refrigerants.

Potential for high efficiency and reduced environmental impact.

6. Phase Change Materials (PCM):

PCMs store and release thermal energy during phase transitions, providing thermal energy storage and load transfer.

It is used in building materials and HVAC systems to improve energy efficiency and comfort.

7. Hybrid HVAC systems:

Combining multiple heating and cooling technologies to optimize performance and efficiency.

Examples include hybrid heat pumps and systems that integrate renewable energy sources with conventional HVAC equipment.

Trends in HVAC Design and Implementation

1. Energy efficiency and sustainability:

Increasing emphasis on designing HVAC systems that minimize energy consumption and environmental impact.

Adoption of energy efficient equipment, advanced control strategies and sustainable practices.

2. Decentralized HVAC systems:

A shift to decentralized systems that provide local heating and cooling.

Benefits include better control, reduced energy losses and flexibility in retrofitting existing buildings.

3. Integration with renewable energy sources:

Incorporating renewable energy sources such as solar, wind and geothermal energy into HVAC systems.

It increases sustainability and reduces dependence on fossil fuels.

4. Focus on indoor air quality (IAQ):

Increasing emphasis on improving IAQ through advanced

filtration, ventilation and air cleaning technologies.

Integrating IAQ monitoring and control into HVAC systems.

5. Electrification of HVAC:

The trend towards the electrification of heating systems, driven by de-carbonisation goals and advances in heat pump technology.

Reduced dependence on fossil fuels and alignment with the adoption of renewable energy sources.

6. Resilient and adaptive HVAC systems:

Design HVAC systems that can adapt to changing environmental conditions and occupancy patterns.

Incorporating climate resistance features to withstand extreme weather events and ensure continuous operation.

Emerging Technologies and Innovations

1. Blockchain technology:

Using blockchain for energy transactions and management in smart grids.

It enables mutual trading of energy and increases

transparency in the use of energy.

2. Nanotechnology in HVAC:

Application of nano-materials for improved insulation, filtration and heat exchange.

Potential for significant advances in energy efficiency and system performance.

3. 3D printing:

Using 3D printing to create your own HVAC components and air ducts.

It reduces production time and costs while enabling innovative designs.

4. Wireless sensor networks:

Deployment of wireless sensors for real-time monitoring and control of HVAC systems.

Improves data acquisition, system diagnostics and adaptive control.

5. Bio-mimicry in HVAC design:

Inspired by natural processes and organisms to develop more efficient and sustainable HVAC solutions.

Examples include passive cooling techniques and energy efficient ventilation systems.

6. Thermoelectric cooling:

Development of thermoelectric materials for efficient cooling in the solid state.

The potential for compact, quiet and environmentally friendly cooling solutions.

Future Directions in HVAC

1. Holistic building design:

Integrate HVAC systems with other building systems (e.g. lighting, security) for a holistic approach to building management.

Use of Building Information Modeling (BIM) for complex design and operation.

2. Personalized climate control:

Development of HVAC systems that provide individual comfort settings for individual users.

Using sensors and adaptive control to create microclimates in shared spaces.

3. Regenerative HVAC systems:

Exploring regenerative design principles to create HVAC systems that have a positive impact on the environment.

Integration of regenerative energy sources and closed systems.

4. Improved Human-Machine Interfaces (HMI):

Improved interfaces for easier interaction with HVAC systems.

Use of augmented reality (AR) and virtual reality (VR) for maintenance and troubleshooting.

5. Decision-making based on data:

Leveraging Big Data Analytics for Informed Decision Making in HVAC Design, Operation, and Maintenance.

Using predictive analytics to anticipate and proactively address system needs.

6. Global cooperation and standards:

Enhanced collaboration between international organizations to develop global standards for HVAC efficiency and sustainability.

Sharing best practices and innovations for industry-wide improvement.